Get to Know Big Cats

CHEETAHS

By Bray Jacobson

Please visit our website, www.garethstevens.com. For a free color catalog of all our high-quality books, call toll free 1-800-542-2595 or fax 1-877-542-2596.

Library of Congress Cataloging-in-Publication Data
Names: Jacobson, Bray, author.
Title: Cheetahs / Bray Jacobson.
Description: Buffalo, New York : Gareth Stevens Publishing, [2024] | Series: Get to know big cats | Includes index. | Audience: Grades K-1
Identifiers: LCCN 2022045086 (print) | LCCN 2022045087 (ebook) | ISBN 9781538285992 (library binding) | ISBN 9781538285985 (paperback) | ISBN 9781538286005 (ebook)
Subjects: LCSH: Cheetah–Juvenile literature.
Classification: LCC QL737.C23 J3365 2024 (print) | LCC QL737.C23 (ebook) | DDC 599.75/9–dc23/eng/20220929
LC record available at https://lccn.loc.gov/2022045086
LC ebook record available at https://lccn.loc.gov/2022045087

First Edition

Published in 2024 by
Gareth Stevens Publishing
2544 Clinton Street
Buffalo, NY 14224

Editor: Kristen Nelson
Designer: Leslie Taylor

Photo credits: Cover Stu Porter/Shutterstock.com; p. 5 ruek66/Shutterstock.com; pp. 7, 24 (fur) Johan Swanepoel/Shutterstock.com; pp. 9, 24 (spots) Sarah Maia Kerr/Shutterstock.com; p. 11 Michal Ninger/Shutterstock.com; p. 13 JonathanC Photography/Shutterstock.com; p. 15 Delbars/Shutterstock.com; p. 17 PETER HATCH/Shutterstock.com; p. 19 Efimova Anna/Shutterstock.com; pp. 21, 24 (cub) Stu Porter/Shutterstock.com; p. 23 Jane Rix/Shutterstock.com.

Printed in the United States of America

CPSIA compliance information: Batch #CSGS24: For further information contact Gareth Stevens at 1-800-542-2595.

Contents

Spot a Cheetah! 4

Fast Cat. 10

Cheetah Life. 14

Cheetah Cubs 20

Words to Know 24

Index. 24

Cheetahs are big cats!

They have short fur.
It is yellow or tan.

They have dark spots.
They have tails
with black rings.

Their legs are long.

They are the fastest land animal!

Most live in Africa.

They live alone or
in small groups.
They hunt in the day.

They make noises.
They chirp and hiss.
They do not roar!

Babies are cubs.
They are born gray.

Mothers teach cubs to hunt.

Words to Know

cub

fur

spots

Index

body, 8, 10
color, 6, 20
cubs, 20, 22
fur, 6, 8